By Gordon L. Ziegler

Child Science

By Gordon L. Ziegler

To order additional copies of this book, contact:
Xlibris LLC
1-888-795-4274
www.Xlibris.com
Orders@Xlibris.com
143750

Contents

Chapter 1

First Things

How many is too many?

There is a law in science (parsimony principle) that things should be made of as few different kinds of elementary particles as possible. Old science (the Standard Model of Physics) requires 61 different kinds of elementary particles to put together light, matter, but not gravitons. New science (the Electrino Fusion Model of Physics) requires only one kind of elementary particle to make light, matter, **and** gravitons. Which do you think is the better science—old science or new science? The one elementary particle in the new science is a negatively charged octon (-O) with 0 spin.

Octal light

The octon is so small, and its charge is so concentrated, that it can ionize octon pairs out of nothing (the hard vacuum). Octon pairs are one negatively charged octon (-O) and one positively charged octon (+O) attracted to each other electrically by the ex-nihilo (out of nothing) force (the meso-electric force). The new octon pairs can have either 0 spin or +1 spin or -1 spin. Unless manipulated electrically or magnetically, the 0 spin octon pairs just recombine again, leaving nothing but an octon of the original charge and the vacuum. Spin +1 octon pairs are just +1 spin octal light, and spin -1 octon pairs are just -1 spin octal light, which travel at the speed of light, though in opposite directions. To conserve charge, spin, and net velocity, bare octons can only ionize two pairs of octons at a time, traveling in opposite directions at the speed of light.

Thus bare octons can generate octal light. Octal light is stable and can travel long distances without decaying.

On collision with matter, octal light provides a source of positively and negatively charged octons, which can be manipulated electrically and magnetically to fuse and form electrinos—the building blocks of light, matter, and gravitons.

Detectable spin

Do you have a toy top or an old phonograph turntable? Can you make them spin fast? If they have no markings on them or dent in them, do they look like they are spinning when they are spinning? No. But if they have markings on them or dent in them, you can see them go round and around when the top or turntable spins.

What is the lesson here? Smooth symmetric devices cannot have detectable spin. This law works also for particles. Smooth symmetric charge distributions cannot have detectable spin. To have detectable spin, charge distributions must be bumpy or lumpy. This law is even more absolute in the particle realm than for tops or turntables. We can make this law one of our system postulates.

In the old science, electrons are thought to be spinning point charges. But that would be smooth symmetric charge distributions, which, according to our postulate, cannot have detectable spin. But electrons have detectable spin. Thus according to our postulate, electrons should not be spinning point charges. Old science must be wrong on this point.

Matching spins and charges

Old science goes by the Quark Hypothesis, which has ⅔ charge and ⅓ charge for the smallest particles to make up everything. They try to match these irregular charges with ½ spins. It is not a very good

2

match. Maybe that is why it takes old science 61 particles to make up light and matter.

New science goes by the Electrino Hypothesis, which has 0 spins for each of the electrinos (1 charge, ½ charge, ¼ charge, and ⅛ charge) and ½ spin for their minimum detectable orbital spins. Doesn't new science have a better match of spins and charges than old science? Maybe that is one reason why new science can make everything out of only one particle.

The smallest particles

In the Electrino Fusion Model of Physics, electrinos are the smallest particles in the Universe. They come in eight varieties—four matter varieties: octons (-O), quartons (+Q), semions (-S), and unitons (+U) (where the signs denote the sign of their charges), and four antimatter varieties: anti-octons (+O), anti-quartons (-Q), anti-semions (+S), and anti-unitons (-U). Octons have ⅛ charge; quartons have ¼ charge; semions have ½ charge; and unitons have 1 charge.

Common particle structures

Pions have four quartons in them—two pairs of orbiting quartons; the pairs orbiting in the opposite direction as the quarton pairs, so that pions have net 0 spin.

Electrons have two semions orbiting in them with either -½ spin or +½ spin. The two semions in the electron sum to a whole charge. Electrons are whole charges. The four quartons in pions sum to a whole charge also, but positive charge.

Neutrons are composed of a +½ spin electron orbiting a 0 spin uniton, with -1 orbital spin—with a net -½ particle spin.

A proton is a pion orbiting a neutron with +1 orbital spin—for a net ½ particle spin. Protons have seven electrinos in them: one uniton, two semions, and four quartons. Neutrons have three electrinos in them: one uniton and two semions.

Positrons are antimatter, and are just oppositely charged (+) electrons (- charge).

One common kind of graviton is just a +½ spin positron orbiting a +½ spin electron, with a +1 orbital spin—for a total net spin of 2. Gravitons can go clear through the earth without colliding with anything.

Shapes and sizes of electrinos

All eight electrinos and anti-electrinos are spherical at rest, but not solid spheres. They each are very thin spherical films of charge. The unitons are the biggest electrinos. Octons are the smallest electrinos. All electrinos are extremely small and imaginary in radius. They cannot be observed even in the most powerful microscope.

Chapter 2

Electrino Fusion

Semion fusion

The simplest electrino fusion is semion fusion. Electrons are composed of two orbiting half charges (orbiting at the speed of light). We call the half charges semions. With the appropriate acceleration (to 940 MeV [million electron volts]), and the flipping of their axial spins, and collisions with oppositely accelerated electrons (to 940 MeV with axial spins), two semions from one electron briefly come into the same orbit as the two semions from the other accelerated electron. The four semions now fuse to two anti-unitons, which are the core particles of anti-protons and anti-neutrons.

These anti-particles scavenge the necessary sub particles from the aether sea to become whole anti-protons or anti-neutrons. These anti-protons and anti-neutrons drift away from the collision point to bump into the brass housing of the collision chamber, annihilating protons and neutrons in the brass in the collision chamber walls. The annihilations produce gamma rays, which can energize solar cells made very efficient by being in the field produced by annihilating positron anti-electrinos (see *Refresher 1 Manual*, Chapter 2 and *Rad Waste-free Power*, Chapter 2).

When semions or anti-semions fuse, the thin film spheres of charge merge into each other, forming a larger spherical thin film of charge.

Switching from matter to antimatter and vice versa

A unique and distinctive phenomenon of electrino fusion is that, whenever there is a fusion of electrinos, there is a switching of the particles from matter to antimatter or vice-versa. This phenomenon enables us to convert whole particles into energy instead of just the mass difference between separate particles and combined particles, called mass defect. This makes electrino fusion reactors 1000 times as efficient as nuclear reactors, and avoids high level radioactive wastes (provided the energy source is in the field of a Refresher).

Clean Energy Source

The above phenomenon can be used to make an electric power plant that is clean—no Carbon emissions, no radioactive wastes, and not even heat pollution.

Chapter 3

Reversing the second law of thermodynamics

Nasty law

There is a law in physics (the second law of thermodynamics) that in a closed system all reactions and energy transfers must tend from more order to more disorder. This law, in most cases, is a nasty law. It is the law which makes people grow old, sick, and die, and their bodies to disintegrate to dust. It is the law which makes it easier to do bad things than good things. Mankind has been exposed to this law ever since the fall of Adam and Eve in the Garden of Eden in the beginning of the world. Before that, there was just the reverse of that law. Everything tended from more disorder to more order. People, animals, plants, and things did not grow old, sick, or die. It was easier to do good things than bad things.

Recently the author discovered how to reverse the second law of thermodynamics from the post-fall state to the pre-fall state using the fusion of positron anti-semions. Originally, and in heaven, God apparently did it through the fusion of octons. That process is instantaneous—"in the twinkling of an eye." 1 Corinthians 15:50-54. The process through the fusion of positron anti-semions is slower and safer for the wicked. The method God used originally through octons puts up with no evil. While it is life to the righteous, it is death to the wicked.

How can we reverse the second law of thermodynamics using positrons? The harmful second law of thermodynamics we now experience is when the arrow between order and disorder points toward the disorder. It is when the change of order energy over time is negative or zero (≤ 0). The beneficial reversed second law of thermodynamics is when the arrow between order and disorder points toward more order. It is when the change in order energy over time is positive (>0). (Order energy is just the positive or negative energy in the creation or annihilation of particles.) It is naturally negative, but we can make it slightly positive by fusing the anti-semions in positrons to unitons (the core particles of protons and neutrons).

It doesn't take much of that to do the trick—10 trillionths of an Amp beam of axial oriented positrons accelerated to 940 MeV in an accelerator and collided with 10 trillionths of an Amp beam inverted axial oriented positrons accelerated to 940 MeV in another accelerator. That would reverse the order to disorder arrow to nearly five acres of land. The effect is backwards to what we would expect. Higher beam currents result in smaller areas affected, and lower beam currents collided result in wider areas affected.

Chapter 4

Miraculous Effects of the Refresher

Reverse aging for adults

The simplest effect of the Refresher to understand is reversing adult aging. Old people can be made young adults again in the active footprint of the Refresher. This effect for positron anti-semion fusion does not really back up time or the clock. It merely reverses the order to disorder arrow in adults. It saturates at the maximum state of order—which is young adulthood. It reverses adult aging at a rate of about 1836 times as fast as the rate the original adult aging occurred. A century of aging can be reversed in just under 20 days.

Resurrections from the dead

The reverse aging occurs also for bodily remains—re-assembling dust and bones into living beings again. All the dead of all ages of earth's history would be resurrected in about 3½ years of Refresher machine time, starting with those who died most recently.

Backing diseases out of existence

In the process of reverse aging, diseases would be backed out of existence. This would work also for difficult diseases like HIV AIDS, cancer, and cystic fibrosis.

Reversing all decay

Spoiled fruit would un-spoil in the active footprint of the Refresher. Fresh fruit would stay at the maximum state of order for fruit forever—fresh picked fruit. And this would be without refrigeration. This would amount to a new kind of food preservation without canning or freezing.

This process would un-decay everything in the Refresher footprint, not just fruit. And the footprint could be enlarged to cover the entire earth.

Reversing pollution out of existence

In the Refresher footprint, all pollution would be reversed out of existence. Depending on the Refresher control settings, this effect could be world-wide.

"Raising up the foundations of many generations" Isaiah 58:12.

The Refresher would automatically rebuild previous decayed structures. It would rebuild and restore the entire earth.

Reversing forest fires

The Refresher not only would stop forest fires in its footprint, but would reverse the fires—restoring all that was lost—animate and inanimate, including lost trees and homes.

Reversing all calamities;

Reversing all effects of war;

Preventing all munitions from firing;

"Making wars to cease to the end of the earth." Psalm 46:9.

Removing sinful propensities from people, including criminals;

Emptying prison houses;

Making possible and efficient Clean Energy Sources.

The blessings of the Refresher are endless. In short it would restore earth to Edenic perfection in about 3½ years of machine time.

www.ingramcontent.com/pod-product-compliance
Lightning Source LLC
Chambersburg PA
CBHW021049180526
45163CB00005B/2344